How Big Is Our Universe?

Mathematical Analogies

by

Dr. Anwar Hamdi

How Big Is Our Universe?
(Mathematical Analogies)

Printed in the United States of America

First Printing: October 2011

ISBN - 978-1466279377

Dedication

I dedicate this book to my family. Without their patience, understanding, and support, the completion of this work would not have been possible.

Table of Contents

Introduction:
Billy of the Universe

Billy grew up in a small remote village where he was born. This town was his entire real and imagined universe! The tiny wooden fence that separates his family's stockyard from their neighbors' was the tallest and strongest fence in Billy's universe. The stream that ran through his garden was the widest and deepest body of water he ever saw! The wooden log that crossed the stream was by far the longest bridge Billy ever walked across.

Without a doubt, the greatest human invention in Billy's eyes was the eight-foot-tall scarecrow his father built to protect their fruit from the birds; and the fastest mode of transportation in town was his horse.

As a teenager, Billy began hearing tales of other towns, even cities, from the older, wiser members of his village. Billy's curiosity, love for adventure, and dreams of discovering new things were all piqued by these stories, motivating his need to travel beyond the confines of his universe. From the beginning of his epic journey, Billy's perceptions changed as his mind expanded.

What was the stockyard fence compared to the Great Wall of China? How did the garden's stream compare to the Nile or Amazon Rivers? How about the wooden bridge he thought was the biggest bridge, did it compare to the Golden Gate Bridge of San Francisco? And would the technical work of his father compare to the pyramids or the Eiffel tower? Even his poor horse, just how fast and strong was he in relation to the sea vessels, trains, metros, and spacecrafts?

Everything in Billy's sheltered reality and limited imagination could only be toys compared to all he now had seen and heard. Then Billy turned from roaming the earth's land and seas to roaming outer space. Once again Billy was overwhelmed by what he heard and saw, things that were hard to imagine. If roaming the earth had made Billy realize his town was insignificant, what he would see in space would make Earth seem equally small.

Ladies and gentlemen, buckle your seatbelts and join Billy on his impressive, surprising, and truly enjoyable journey.

Point of takeoff: Planet Earth

Date of travel: Every turn of the page

An important message before takeoff: All comparisons made in this book are true and very accurate, with the reference numbers included for each.

Chapter 1: EARTH

Lebanon and Earth.....

Earth, this magnificent planet we live on, is BIG.
Very, very big. Its surface area [1]—including the seas
and oceans, is unbelievable. It is large enough to fit
49,000 countries the size of Lebanon [2].

Take a plain sheet of paper with 25 lines. On each
line write the name of one of these 49,000 countries.
You will need 1960 sheets of paper to accomplish
this task. That is a total of four binders, each binder
with 490 pages.

If writing each of those countries' names takes only
five seconds, you will need to write nonstop, day and
night, for three days!

==

1. Earth's diameter: 12760 km (7929 mi.).

2. Surface area of Lebanon 10452 km^2 (4035 $mi.^2$)

How many of Khufu's pyramids can we fit inside of our planet Earth?

Khufu's pyramid (the biggest of the Egyptian pyramids) along with its brothers Khafre and Menkaure, is one of the Seven Wonders of the World.

It is said that 100,000 laborers worked on Khufu's pyramid for 20 years! The length of its square base is approximately 230 meters (755 ft), and its original height estimated at 147 meters (482 ft).

The size of this pyramid is equal to a large 20-story building, 200 meters (656.2 ft) in length, 130 meters (426.5 ft) in width, with the height of each story being 5 meters (16.4 ft)!

Our planet Earth can fit more than 419 million, million of Khufu's pyramid.

If Khufu's pyramid took 100,000 laborers 20 years to build, how many years would it take just to fill our planet with these premade pyramids? Imagine that you have these premade pyramids, each the size of Khufu's. Now, imagine that the planet Earth is an empty sphere and that you have monster-sized cranes that can lift one of these pyramids and place it into the earth's sphere every second.

Now imagine you have 100,000 of these Monster cranes all working at the same time.

In order to fill the earth with these pyramids, you would need to monitor these 100,000 cranes each working at a speed of 1 pyramid/sec for over 132 years!

Have you started to marvel at the size of our Earth yet?

Can we weigh Planet Earth?

Imagine if you had a very, very large balance scale and you placed planet Earth in one hand and all of the people now living on the face of the earth, which are approximately 6.875 billion [1], on the other hand. Let's say that each one person, including the kids, babies, elderly, etc. weighed 100 kg (220 lb.). That makes everyone a heavyweight! Would these people have enough mass to balance the earth's weight [2]?

If you considered all these 6.875 billion people as one group, you would need not one, not two, not three, but about 9 million, millions of these groups to balance the earth's weight.

That's 8,692,364,000,000 groups of 6,874,963,746 (6.875 billion) humans, each weighing 100 kg (220 lb.)!

Can't imagine that? I'll help.

Imaging that every second you loaded one of these groups on to the scale. Meaning, every second you would add 6.875 billion people weighing 220 lbs. each.

Q: In order to balance the earth's weight, you would have to work continuously for…can you guess?
A: More than 2756 centuries!

Since the birth of Christ, it has only been 21 centuries.

Did that help? I don't think so. I think we both need some help trying to imagine that.

Let's try again, shall we?

Remember Khufu's pyramid that we talked about earlier? It took approximately 2.3 million stone blocks to build it, each block weighing in at an enormous 2.5 tons!

Taking that into account, the weight of the entire human population with each person weighing 100 kg (220 lb.) compared to the weight of the earth is similar to approximately 10 grains of wheat [3] compared to the weight of Khufu's pyramid!

So next time you visit Khufu's pyramid in Egypt, place 10 grains of wheat next to it and imagine that this is the entire human population's weight compared to the weight of our planet Earth!

==

1. According to the USA census, the world's population was 6,874,963,746 people in 10/14/2010.

2. Weight (mass) of planet Earth: 5.976×10^{24} Kg

3. 1 grain of wheat = 64.79891 mg

Around the world in 83 Days

Imagine you were in an excellent physical shape, a super human running machine that could cover 20 kilometers (12.4 mi) an hour. Have you ever tried running that fast?

Now, imagine that it would take you a little over 83 days of nonstop running at this exhausting speed to run all the way around the world!

The Incredible bullet!

Have you ever seen anyone fire a powerful gun? I would like you to observe that one day.

When you do, see if you can track the bullet just as it exits the gun's barrel, then try to catch it before it reaches its target!

Indeed, this is only something that Superman would be capable of. Bullets leave firearms traveling at a speed far greater than one can see with the blind eye, let alone attempt to catch with our bare hands. With this picture in mind, try to imagine that our planet Earth orbits the sun at a speed 60 times faster than that of a bullet!

The incredible speed of a bullet compared to Earth's orbit speed is the same as a man running compared to a fighter jet above him traveling at supersonic speed.

Earth, the spinner top!

Our planet Earth, with its enormous size, spins around itself with a tilt at a breathtaking speed!

Disastrous hurricanes are known to have very fast wind speeds. Hurricane Andrew, for example, caused devastation to no less than 90,000 US homes in 1992, with recorded wind speeds of 270 kilometers/hr (168 mph).

Throwing and destroying all that crosses their path, these fast winds leave only ruins to be remembered.

Our Planet Earth revolves around itself at a speed faster than the eye wall of the fastest (480 kilometers/hour) of hurricanes by 3.5 times!

Chapter 2: The Moon

Closest of neighbors:

A long and exhausting plane trip from New York to Rome, takes approximately nine hours nonstop. During this trip the plane covers a whopping distance of 6,891 kilometers (4,283 miles). If you were to redirect your plane toward the moon as a final destination, you would need to brace yourself for a nonstop 21-day trip!

The humble neighbor:

The moon's circumference roughly equals the distance from New York to Cairo (Egypt).

The moon's atmosphere is very small—so small that all of it could be collected and stored in a small house. Even then, its pressure would not equal one atmospheric pressure on Earth.

Parts of Earth are bigger than the moon:

The combined surface areas of only Russia, Canada, the United States, and Chad are more than that of the moon [1]!

==

1. Diameter of the moon: 3475 km (2159 mi.)

Where Men turn into children!

A full-grown man weighing in at 75 kg (165 lb.) only weighs about 12 kg (26.5 lb.) on the moon [1]. That's less than a small child.

Race the moon?

The speed of the moon in its orbit around the earth is faster than an F-15 fighter jet [2]!

In fact the moon is at least three times faster than the speed of sound [3]!

===

1. Gravity on the moon's surface is 16% that of Earth's.

2. An F-15 fighter jet can reach speeds of 2.5 Mach (sound).

3. Speed of sound is 1223 km/hr (depending on the type of medium it travels in)

Chapter 3: Mercury

The cockroach and the jet plane!

Mercury is the closest planet to the sun [1]. It orbits around the sun at an incredible speed [2]—faster than a 747 Boeing plane by 195 times [3]!
What would you think about a race between a cockroach [4], and a jet plane moving at a speed faster than 1000 kilometers (621.4 mi.) an hour?

That is a similar analogy to a race between a Boeing 747 and the planet Mercury.

=====================================

1. Distance between Mercury and the sun is approximately 57.9 million km (36 million mi.).

2. Mercury solar year is 88 days.

3. Speed of a Boeing 747: 885 km/hr (549.9 mph).

4. Speed of a cockroach is approximately 1.5 m/sec

Chapter 4: Venus

Where a day is longer than a year!

On Earth, a year is the time required for the planet to complete one circle around the sun. A day is the time our planet takes to revolve around itself. A year is greater than a day by 365.3 times. But on Venus, a day is longer than a year!

Venus revolves around itself at such a low speed, meanwhile circling the sun [1] at outstanding speeds [2]. In fact, Venus's orbit speed around the sun is 19,000 times greater than its speed of revolving around itself[3,4]!

So, what if our planet was similar to Venus in its orbit pattern?

If our year was still 365 days, each day would be 300 days in length!

If our days were still 24 hours, then each year would only be 29 hours! That means each new year would be a little less than a day and a half away.

We would all be very old, but at least a New Year's resolution is always right around the corner!

==

1. Distance of Venus from the sun: 108 million km (67.1 million mi.)

2. Venus's year: 225 Earth days

3. Venus's day: -243 Earth days (negative because it rotates the opposite direction of all the other planets)

4. Diameter of Venus: 12106 km (7522.32 mi)

Chapter 5: Mars

The largest volcano in our solar system

On the red planet we call Mars, there is a dormant volcano that is millions of years old. It is believed that this volcano is the largest volcano in our entire solar system.

The top opening of this volcano is three times higher than Mt. Everest [1], the highest point on Earth! Its base is estimated to be 600 kilometers (372.82 mi.) wide!

A valley deeper than the ocean

There is a valley on the surface of Mars that is approximately 4000 kilometers (2485.5 mi.) in length, 250 kilometers (155.3 mi.) in width, and a depth [2] reaching 4000 (2.5 mi.) meters!

===

1. Highest point on Mt Everest: approximately 8,848 m (4.971 mi.)

2. Average depth of the ocean: 3,794 m (2.4 mi.) deep

Chapter 6: Jupiter

The earth's stamp on the Jupiter envelope

Take a stamp 2 cm (0.787 in.) in length and 1 cm (0.393 in.) in width and place it in the corner of a 25 cm (9.842 in.) by 10 cm (3.937 in.) envelope. Then observe. This is what the earth's surface area is like compared to Jupiter's [1].

In fact, you can copy and paste 125 Earths on to the face of Jupiter. And what is inside Jupiter is even larger.

Jupiter can hold 1400 planet Earths inside of its belly!

Gases heavier than dirt!

Even though Jupiter is primarily made up of helium and hydrogen gas, it outweighs the earth by 315 times [2]. This is due to its colossal size.

==

1. Diameter of Jupiter: 142,800 km (88731.8063 mi.)

2. Jupiter's weight (mass): 19×10^{26} kg

The giant spinner

If you wanted to embark on a running trip around Jupiter like the one you did around the earth, then you would need to brace yourself for a very long trip—a trip longer than the distance between the earth and the moon!

If your trip around Earth was 83 days at a nonstop running speed, this trip would take you 2.5 years!

This giant also revolves around its axis [1] at a great speed—a speed greater than the earth's rotational speed by 27 times!

===
1. Jupiter's day: 9 hrs and 55 minutes Earth time

Chapter 7: Saturn

The floating Planet!

Saturn is heavier than the earth [1] by 95 times! But due to its colossal size [2] (835 times the size of Earth,) its density is less than that of water [3].

If there were a swimming pool large enough to fit Saturn, Saturn would float!

Jewelry no one can afford

Saturn has rings than spin around it. Like necklaces, these rings bestow Saturn with beauty and magic. These rings are indeed very large and as necklaces, would be quite expensive. For example, if you were to take a gold necklace of only 1 mm in thickness, its width would have to be at least 60 meters to be relatively comparable in thickness and width to Saturn's rings [4].

===

1. Earth's density: 5.5 g/cm^3

2. Saturn's diameter: 120,660 km

3. Density of water: 1 g/cm^3

4. Width of one of Saturn's rings can reach up to 300,000 km, although it is only a few kilometers thick.

Chapter 8: Uranus

The chicken egg, the ostrich egg, and the grain of sand!

The average distance of Saturn from the sun is over 1,427 million kilometers (886,696,691 mi.). This distance was the old border of our solar system. When Uranus was discovered in March 1781, our solar system's size doubled! This is because Uranus's average distance from the sun is over 2,870 million kilometers (1,783,335,321 mi.)!

So our solar system went from being the size of a chicken egg[1] to the size of an ostrich egg [2], even though all that was added to it (Uranus [3]) does not compare in size to a grain of sand [4] on a beach 110 meters (360.89 ft) long!

===

1. Length of a chicken egg: 7 cm

2. Length of an ostrich egg: 15 cm

3. Diameter of Uranus: 52,400 km (32,559.8 mi.)

4. I assumed each a grain of sand to have a diameter of 1 mm (this size could actually contain a few grains of sand).

Chapter 9: Neptune

From previous studies on our solar system, mathematical and astronomical, it was postulated that there must exist an eighth planet in order for balanced equilibrium to exist. From there on, calculations were made on this unknown object.

In 1846 A.D, 65 years after the discovery of the last planet (Uranus), this unknown object—planet Neptune—was found almost exactly as anticipated by previous calculations.

Chapter 10: Pluto

Toy ship in the Nile

Say you were to build a small toy ship, about 80 cm in length. Then you would place this little ship at the beginning of the Nile River [1] and let it sail away.

You would watch your ship float down the river with elegance, ease, and painfully slow progress. Say it moved at a rate of 73 meters (239.5 ft) a day; when would you expect your boat to sail over the waters of the Mediterranean Sea?

Maybe your great-great-great-great-great-grand children would witness that event. That's because your boat would be expected to arrive after 248 years.

This is similar to how long it takes Pluto (the dwarf planet [2]) in its journey around the sun [3]!

===

1. Length of the Nile: 6,656 km (4135.8 mi.)

2. Diameter of Pluto: 2200 km (1367 mi.)

3. Average distance of Pluto from Sun: 5,900 million km (3,666,090,034 mi.)

Chapter 11: Sedna

One year equal to more than ten thousand Earth years

The time it takes Sedna to complete one revolution around the sun (one year) is enough time for our planet Earth to have completed 10,500 revolutions around the sun (10,500 years)!

Smaller than Pluto, double the size of our solar system!

As previously mentioned, when Uranus was discovered in 1781, our solar system doubled in size. When talks about a new "dwarf planet" (Pluto) surfaced, our solar system doubled in size once more.

When Pluto was discovered in January 1930, our solar system's length expanded to roughly 11,800 million kilometers (7,332 million mi.). Most people simply dismissed the thought of our solar system's borders being expanding any further than that.

Yet once again, our solar system's dimensions doubled when Sedna was discovered on November 14, 2003, as Sedna spans over 11,500 million kilometers away from the sun.

What do all of these dimensions mean to you?

Do you recall our plane trip from New York to Rome? The one where we decided to redirect the course of the trip toward the moon? Yes, that trip that had us on the plane for 21 days, remember?

Now let's try to reroute our plane towards Sedna. How many more days do you think this trip will take?

This trip, ladies and gentlemen, will take us a total of 17 centuries. That's right, it will take us a total of 1,700 years of nonstop travel in our Boeing 747. Your great-great-great-grandchildren would never reach it. Neither would your 20th or even 40th generation of grandchildren dream of seeing Sedna. It is obvious that our Boeing plane is unequipped for such a trip.

This trip requires a much better and faster aircraft than the one we have now. Let us then buy tickets aboard a space shuttle.
In order for a space shuttle to "break loose" from the earth's gravitational force, it must travel at a very high takeoff speed exceeding 40,000 kilometers (24,854.8 mi.) an hour! At this speed, a space shuttle can circle the earth at the equator in 1 hour. It can also reach the moon in under 10 hours!

Let's say, hypothetically, that our space shuttle is able to maintain this magnificent takeoff speed for the entire trip. It would still take our shuttle 81 days just to reach our closest neighbor, Mars [1]!

If we decided to continue to Sedna, the current border of our solar system, that would extend our trip by some time. In fact, you might have aged quite a bit before you reached your destination, considering it would take you a total of 34 years of nonstop travel.

===

1. Mars's distance from Earth: 78 million km (48,466,953 mi.) *at its closest distance

Snail in Dubai crawls to Rome

It is obvious by now that until we develop a more capable aircraft, space travel—even within the realm of our neighborhood (our solar system)—is completely impractical.

It is exactly as follows:

A small snail in your backyard travels at an average speed of 2.5 mm a second, assuming you clear its path of any obstacles. Now place this snail in the center of Dubai (United Arab Emirates) and clear its path. With all the incentive necessary to ensure your snail heads in the right direction, you will, in your very, very fast space shuttle described above, arrive at Sedna in roughly the same time it takes your snail to slither its way from Dubai to Rome!

Chapter 12: The Sun

A "Sun day" equal to a "lunar month"

The time required for a planet or star to complete one revolution around its axis is called a day. It takes the Sun a total of 27 Earth days to complete one of its own days [1], even though it rotates at a speed faster than a Boeing 747 [2] by seven times!

Smaller than a sesame seed on a loaf of bread

Draw a circle with its radius equal to 10.91 cm (4.29 in.). Draw another circle with a radius of 2 mm (0.0787 in.), inside the circle you drew previously. The first circle is a representation of the sun's surface area [3], and the second is the earth's surface area [4]!

Now, with a pen, quickly make a dot the same size as that of the earth in the previously described paragraph. Repeat this process to make a dot every second. It will take you 1 hour and forty minutes just to fill in half of the sun's surface area!

This is how our planet's surface area relates to that of the sun.

Earth curtains to block the Sun!

If you were to place curtains around the sun, with each curtain the size of the entire earth's surface, you would need 12,000 curtains! If hanging each one of these curtains only takes five minutes each, then the last curtain in this play of "blocking the sun" would be hung after 41 days of nonstop curtain hanging!

===

1. The sun revolves around its axis once every 27.3 days. .

2. Boeing airplane 747 speed: 885 km/hour

3. Diameter of the sun: 1.392 million km (864,948.69 mi.)

4. Diameter of the earth 12,760 km (7,928.69 mi.)

A lentils-stuffed watermelon!

Go to your local produce market and shop around for a large, round watermelon. Look for a one with a diameter about 55 cm (21.65 in.). It will probably be expensive considering its size, and you might have some difficulty lifting it, but go ahead and purchase it. Now, buy a bag of lentils and take only one seed out. The diameter of a lentil seed is about half a centimeter, the size of this circle drawn at the end of this sentence. ◯

The size of Planet Earth compared to the Sun is about the size of that lentil compared to the watermelon.

Say you were to gut out the watermelon to a very thin crust and place the lentil seeds inside your hollow watermelon. If you were to fill up your watermelon entirely with lentil seeds, you would need 1.3 million lentil seeds.

If you started filling your watermelon with lentils on Monday morning at 8 a.m. at a rate of 100 seeds a minute (a little more than one seed a second), it would take you until 8 a.m. Wednesday morning of the next week to accomplish this task. In other words, nine days of continuous stuffing!

This is how our earth's very large size compares to the sun!

The 14-year-old child

How beautiful and radiant is our sun?! It is still in the beginning of its youth and development, as its age is only about 5,000 million years old. In star years this is about the age of young teenager.

Who can imagine this age?

Let's make an attempt, shall we? Let's say we hold the fast-forward button on a time device so that every 10 years that pass only take 1 minute. It would take this device only 6 to 7 minutes to fast forward through most of our lives, while it would take 950 years to fast forward through the life of the sun.

Let's try to simplify it even further.

Your eye blinks every 2 to 10 seconds. The blink itself only lasts about 0.3 to 0.4 seconds. By the time your son or daughter reaches eight years of age, his or her eyes will have blinked millions of times. Imagine now that your child's age compared to the age of the sun is equal to just one blink during their eight years of vibrancy and life!

Imagine eight years of human life, full of events, stories and adventures compared to a blink of an eye.

The mouse and the elephant

Have you ever seen a great African elephant, on TV or in real life? Its weight can reach up to 7 tons (15,680 pounds). Yet, in cartoon form he is portrayed as being scared of a little mouse. Two small mice may not weigh 1.5 ounces combined.

How many little mice does it take to achieve an elephant's weight? Our planet Earth compared to the sun is the same as the little mouse compared to the elephant!

If you were to place the sun on one side of a scale, you would need 333,000 Earths on the other side to balance it out. If you had a hard working imagination that could place 33 planet Earths on the scale every minute, you would be working on this project for an entire week!

Running of the Sun

Our planet swims through space at a magnificent speed that reaches 100,000 kilometers an hour! That's far faster than any rocket or space shuttle currently available. Yet even at this remarkable speed, it takes our Earth 365 days to accomplish one revolution around the sun. That is just how long its track is!

On the other hand, our Sun's situation is even more enticing and mysterious.

Our Sun runs through space at a speed much faster than our Earth does! In fact, it does so at a speed of 900,000 kilometers an hour. So if you were to place the sun in the earth's track, it would only take the sun less than 44 days to complete one revolution. However, the Sun has a track of its own—one, which is much, much longer than that of the earth's. Even with the Sun's outstanding speed, it takes about 200 million years for the sun to complete one revolution around the center that it revolves around in the galaxy.

That's right: 200,000,000 years! No typo there! According to our understanding, one Sun year is equal to 200 million of our Earth years.

So Just how long is the sun's orbit then?

That would be 1,578,096,000,000,000,000 kilometers!

Do you remember our snail that crawled at a speed of 9 meters per hour? Do you also remember our space shuttle that maintained the incredible takeoff speed of 40,000 kilometers/hr. while trying to escape Earth's gravitational field?

If you were to climb aboard this space shuttle in an attempt to complete just one revolution of the sun's orbit mentioned above, how long might this take? You—I mean maybe your 100 millionth grandchild, would complete this journey when the descendants of our snail friend finish crawling from the earth to the Sun and back, and repeat this cycle over 1,180 times!

Unbelievable isn't it?

Is it a nuclear oven, a nuclear reactor, or what exactly?

When the atomic bomb was released on August 6, 1945, the entire world realized the power of the atom. This atomic bomb was equal in force to 20,000 tons of the extremely explosive TNT. Keep in mind that the amount of uranium substance used in generating such a large destructive force was a little less than a handful!

This was the atomic bomb. The hydrogen bomb (nuclear bomb) on the other hand was able to generate energy much stronger and more capable than that of the atomic bomb. Indeed, the energy released from nuclear fusion or fission is far greater than that of atom splitting.

On the first of November 1952, in its first trial, the hydrogen bomb was able to generate a force equal in magnitude to 10 million tons of explosive TNT. It was 500 times stronger than the first atomic bomb.

The energy released from the fusion of two small atoms of hydrogen to produce one larger atom of helium is enormous and beyond comprehension. It was found, according to Einstein's equation, that the energy released is equal to the mass (which will be transformed to energy) multiplied by the speed of light squared! (The speed of light is 300,000 kilometers/ sec; 186,411.358 mi/sec).

On this basis, if 1 kg (2.2 lb.) of mass is transformed

to energy, that energy released would be equal in magnitude to the explosive force generated from 22 million tons of TNT.

Inside the Sun, at a depth of about 650,000 kilometers, about 1.5 times the distance from the earth to the moon, nuclear reactions occur that make all of current nuclear weapons, without exaggeration—seem like child's play!

Inside the belly of the Sun, pairs of hydrogen atoms are constantly fusing to form helium. According to the laws of nuclear fusion energy is released!

The energy released from the first atomic bomb was unbelievable and was equal to only 20,000 tons of TNT. So what would you think of an energy equal to 22 million tons of TNT?

Better yet, what would you think of energy equal to the explosion of 22,000 million tons of TNT?

Well, for starters, it would be capable of wiping out the entire human population fourteen times over—in addition to completely incinerating every city, town, and for that matter every structure on the face of the earth to oblivion!

Now what would you think of not just transforming 1 ton of mass into energy, but 6,000,000 tons of mass all at once?! It would give every human on the face of the earth their fair share of 1,100 atomic bombs like the one used in 1945— EVERY SECOND!

In the Sun, every second around the clock, 6,000,000 tons of mass is transformed into energy, which is transmitted into space.

Based on that, the Sun's mass contracts 6 million tons every second. So every second, the sun gets lighter by a weight equal to one Khufu pyramid! And every 30 million years, the sun loses weight (mass) equal to that of the earth!

Very little equals very much!

The energy liberated from the sun is equal to 380 million, million, million, million watts! Every Second! From what is described as the Sun's glow or glare, the earth receives about 180 thousand, million, million watts of energy every second!

The energy expenditure of the entire human race is estimated at about 8 million, million watts every second. This means that the sun gives humanity 22,000 times its need for energy!

The Sun is similar to a very, very wealthy man, like a billionaire if you will, with more than 2,000 million dollars! When you ask this person for help, he digs into his pocket and gives you one dollar. How odd! This is what our relationship with the sun is. Yet even this insignificant amount of energy, relative to the sun, is far beyond what we need.

Assume that your monthly expenditure is one

thousand dollars. You are able to live very comfortable with this amount of money, yet your boss offers you a raise to 22 million dollars! Not even in your best dreams. This is how generous the Sun is with its energy reaching the earth.

Tongues of the sun

Our huge sun is exploding and boiling. The boiling is a little difficult to comprehend.

Inside the sun, temperature reaches approximately 15 million degrees Celsius. On the Sun's surface, the temperature is estimated at about 6,000 degrees Celsius. At the sun's core, the temperature can reach 300 million degrees Celsius.

Keep in mind that water boils at a temperature of 100 degrees Celsius. Iron melts at a temperature of 1,430 Celsius!! So can you begin to imagine how hot the sun's surface is at 6,000 degrees Celsius? Forget the sun's core, which has temperature of up to 300 million degrees. Everything you know about the laws of physics, atoms, motion, shapes, etc. becomes an entirely different world.

During all of this burning and boiling, the sun expresses "tongues" of fire in the form of heat projections or fire fountains. These fire projections reach far into space, to a distance about half of that between the earth and the moon! The circumference of each of these "fire fountains" or projections is bigger than the earth's circumference by 15 times!

Chapter 13:
The Solar System

A trip through the solar system

The Sun, the planets that revolve around it and their moons, and everything in between, including space objects and space dust, are collectively called the Solar System.
Our solar system is at least 24 thousand million kilometers long! That's 160 times the distance from the earth to the Sun!

Our Space shuttle that could circle the earth in one hour required five months to reach the sun. In order for this shuttle to travel from one end of our solar system to the other, it would take 70 years!

The cockroach and the car!

Our solar system, with its colossal size, moves, or "swims" in space!

We were amazed by the Boeing 747, such a huge plane that travels at such fast speed. So what do you think of our solar system, with its astronomical size, traveling or "swimming" at a speed 77 times greater than that of a Boeing 747?

An F-15 fighter jet's speed, which is both exhilarating

and scary both at the same time, is slower than our solar system by 23 times!

The speed of this super jet compared to the speed of the solar system [1] is equal to a cockroach [2] entering a race against your car at speed of 120 kilometers per hour.

Living in a vacuum

If you were to line up the sun and all of the planets and moons in our solar system back-to-back, the total length of this "planet chain" would not exceed 2 million kilometers. Keep in mind that our entire solar system's length is at least 24 thousand million kilometers.

Enter a large ballroom that is 50 meters long and 20 meters wide with a ceiling 10 meters high! In this huge, empty room, look in the corners or on the wall for a small spider that is no bigger than 2 millimeters (smaller than a sesame seed). All of our solar system's planets and moons combined occupy the same space relative to our solar systems as the small spider does inside that large ballroom.

===

1. Speed of our solar system: 19.3 km/sec.

2. Cockroach's speed: 1.5 m/sec.

The planets' designs

It is commonly thought that all planets orbit the sun in simple oval-shaped tracks. However, tracking these planets has revealed some amazing designs and sketches that are unbelievable!

The following pictures are the orbits of some of our planets around the sun, as seen from Earth.

Venus's Path

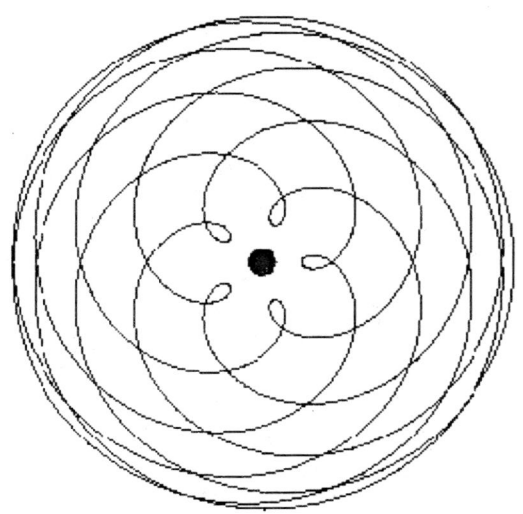

Mercury's path

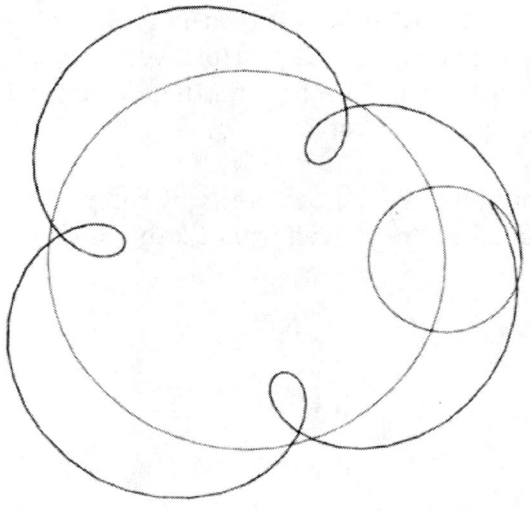

Mars's path

Jupiter's path

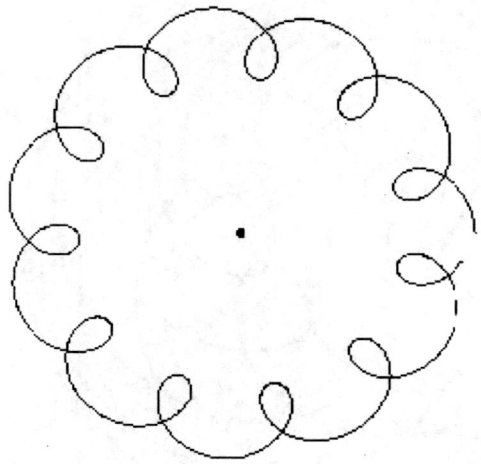

Chapter 14: The Galaxy "The Milky Way"

Your astronomic address

If you were to ask me about my Earth address, I would say for example: Baton Rouge, Louisiana, United States. If you were to ask me about my astronomic address, I would say I live in the American continent on Earth in the solar system within the Milky Way galaxy.

The same way Baton Rouge is one of the many cities in United States, our solar system is one of 200,000 million other solar (star) systems in the Milky Way!

Can you describe the amount of joy all parents have when their first child goes to his or her first day of school? As you know, this child will only be one of many millions of students currently in schools, colleges, or post-graduate institutions. This is the situation with our solar system relative to other solar systems currently in the Milky Way.

Indeed, our one and only Sun is just one of 200,000 million stars in our galaxy.

Just as our Sun has formed its own solar system with many planets revolving around it, the rest of the 200,000 million stars in our galaxy are believed to have formed their own solar systems as well. This means that our galaxy contains over 200,000 million solar systems!!

One-million-dollar prize for he/she who can count the stars!!

Assume you had the tools to see all of the stars in our galaxy! Could you count them all? What if there was a prize involved? Our only rule would be that you must count them yourself, without the aid of anyone or any electronic devices. You see a star and you count it. Simple, right? Let's also assume that you are fairly fast at counting, so you count one star, two stars, three stars, etc. at a rate of two hundred stars a minute. Once you have counted all of the stars in the galaxy, you will receive a one-million-dollar prize! Wouldn't you be excited to participate in such a game? This game would rely on time and patience, not intelligence, so even if you were a math teacher or genius you wouldn't be at any special advantage.

In order to count the stars in our galaxy only, not all the stars in the sky, you will need to work hard and nonstop without boredom for longer than 1,900 years!

That's right: just to count the stars in our galaxy, not including the planets, will take you (or should we say your great- great- great- fill- in- the- blank- grand children) over 19 centuries. That's a lot of time.

Measuring distance with time

When we pass the borders of our solar system, measuring distances with kilometers or even millions of kilometers is simply meaningless. To estimate these astronomical distances, we need a new unit of measurement. This unit is "the light-year" [1].

Light, you see, travels at an incredibly great speed. In fact, it is the fastest thing currently known to exist. Light can travel a distance of 300,000 kilometers every second.

Do you remember our very fast spacecraft that could maintain a speed of up to 40,000 kilometers an hour—that same spacecraft that could take you to the moon in 10 hours? Well, light can reach the moon in 1 ⅓ seconds flat!

Our very fast spacecraft compared to light is the same as a snail [2], moving at 9 meters per hour (0.0056 mph), racing against you in your Porsche, BMW, or whatever vehicle travels at a speed of 240 kilometers an hour (149 mph).

Looking for a needle in a haystack

These 200,000 million stars in our Milky Way are scattered throughout the galaxy so that the average distance between any two stars is at least 5 light-years!
The first star we encounter on our way upon leaving is, of course, the Sun. The Sun is approximately 8 "light minutes" from the earth. If we were to embark on a trip to the nearest star (other than the sun,) it would be a very long trip. The first star we'd encounter would be at a distance greater than 40 million, million kilometers. That is more than four light-years!

If the distance between Earth and the sun was equal to the distance between your front door and the side walk (roughly 10 meters,) then the distance from Earth to the next star would be greater than the distance between Los Angeles and New Orleans [3].

On a stretch of 17 light-years from Earth, there are only 45 stars total. Let's simplify this.

Every time you cover a distance of 600 kilometers with your spacecraft, place a black flag on the side of the road. And every time you encounter a star, place a red flag on the side of the road. At the end of your 17-light-year trip, you will find that you have placed more than 265,000 million black flags and only 45 red flags. That's 6,000 million black flags for every red flag. This is like trying to pick out a specific person from the entire world population!

That's a lot harder than finding a needle in a haystack, don't you think?

===

1. A light-year is the distance light can travel in a year: more than 9 trillion km

2. Snail's speed: 2.5 mm/sec

3. Distance between Los Angeles, California and New Orleans, Louisiana is 2,725 kilometers.

The Milky Way and fried eggs

Our galaxy the Milky Way is flat and disk-like, similar in shape to a lentil seed or two fried eggs placed back-to-back.

This lentil seed or fried egg would be inconceivably large. Words like huge, large, colossal, or enormous lose meaning when trying to effectively describe or help the reader imagine the true dimensions of our galaxy.

Our galaxy's thickness is no less than 20,000 light-years! Its width is 100,000 light-years and its length is estimated at over 400,000 light-years!

The length of our solar system is no less than 24,000 million kilometers. This huge number is only about 22 light hours, according to our new measuring stick.

Our galaxy, on the other hand, is 400,000 light-years long!

If you were to scale down our solar system into a 2-centimeter-long string, then our galaxy would have to be a string that started from Riyadh, Saudi Arabia, and ended at London Heathrow International Airport, United Kingdom [1]!

Our galaxy is longer than our solar system by 160 million times! Light, which can travel the estimated 150 million kilometers between the earth and the sun in about 8 minutes, needs 400 thousand years to cover the distance from one end of our Milky Way to the other.

<hr>

1. Distance between Riyadh (Saudi Arabia) and London (UK) is about 3074 km (1,893.3 mi) by air.

A Fly in front of the pyramid!

Once again, stand in front of the great Khufu's pyramid in Egypt. While you begin to admire and appreciate this massive mountain of rock, a fly passes by your ear, so you wonder: just how would this small fly [1] compare to this pyramid?

There is really no room for comparison, is there? Although, I have asked myself an even tougher question: how would this fly compare to not just one but 2,900 pyramids, each the size of the great Khufu's pyramid? The answer is "not anything"!

The weight of the entire planet Earth (mass) compared to the mass of our Milky Way galaxy is similar to that fly in front of 2,900 large pyramids!!

==

1. Fly's weight: 25 mg

Variety of stars

Our galaxy contains 200,000 million stars. These stars come in a variety of shapes and other characteristics, which give our galaxy the appearance of a colorful garden.

Some stars make our sun look tiny. For example, the star "Eta Carinae," which is at least 8,000 light-years away from Earth, has a mass 100 times larger than our sun. Its shine or glare is at least 6 million times greater than that of our Sun. Another big star is "Betelgeuse," which has a diameter 800 times longer than our sun.

Pulsar and neutron stars

We have, in our solar system, so-called "pulsar" or "neutron" stars. The diameter of these stars does not exceed 20 kilometers on average; remember that the diameter of the sun was 1,392,000 kilometers. On the other hand, a piece of one of these stars no larger than the tip of a needle weighs more than one million tons [1]. In these stars, the space between electrons and protons that make up the atom is eliminated, and electrons and protons fuse.

These very heavy stars spin or revolve around themselves at incredible speeds. For example, the star PSR 1937+215 spins around itself at more than 55 million times a day [2], whereas our Earth completes only one revolution per day.

The current number of pulsar stars in our galaxy only is no less than 50 million. I said "current" because a new star forms every 50 to 100 years—so be on the lookout.

The temperature on the surface of our sun, as we said previously, is about 6,000 degrees Celsius. The temperature on the surface of one of these pulsar stars is estimated at 10 million degrees Celsius. This means that the surface temperature on these neutron stars is 1,666 times greater than the surface temperature of the Sun!

==

1. Density of Pulsar star: 5.5 million tons/cm^3

2. PSR 1937+215 revolves around itself once every 1.557 milliseconds.

Black holes!

According to studies, it is estimated that in our galaxy there are at least 100 million black holes caused by stars collapsing!

Black holes are very, very large stars (usually much larger than our sun) that in their final stages of life, enter what is described as the collapsing phase. The star becomes very dense, so dense in fact, that even light cannot escape its gravitational force, hence the name.

The black hole's density becomes infinity while it is surrounded by an area of several kilometers. Nothing can escape its grasp!

If our sun completes its collapsing phase, the resulting black hole would have a diameter of only 3 kilometers.

Here, of course, all the laws of physics as we know them become obsolete. The new laws governing this phenomenon are a complete mystery to human knowledge and even to our imagination.

It is believed that there is a black hole at the center of our galaxy, resulting from the collapse of a large star four million times the size of our Sun!

Just for comparison, in the galaxy named M87 there is a black hole with a mass 3,500 million times the mass of our sun!

The entire human race fails to lift one drop of water!

A star the same size as our sun, with a diameter of about 1,392,000 kilometers, must collapse or "shrink" to a size of only 3 kilometers in diameter for it to become a black hole. Even with this incredible shrinkage in size, the weight remains the same!

To paint the picture even better, imagine that our moon [1] was compressed to the same extent as that of a black hole; it would become the size of water necessary to fill a medium-sized pool 11 meters (36.08 ft) long, 10 meters (32.8 ft) wide, and only 2 meters (6.56 ft) deep, though its weight would remain the same.

So if you were to take one drop of this water, about 1 mm^3, its weight would be 333 million tons!

If every human being on the face of the earth was able to lift a 50 kg weight, we, collectively, would not be able to lift one drop of this water!

==

1. Mass of the moon: 0.0123 of the earth's mass.

To the moon in ten minutes

Our vast galaxy, with its 200 thousand million star systems, which span over a length of 400,000 light-years, moves or swims in space at an incredible speed!

It can cover the distance between the earth and the moon in less than 11 minutes!

Chapter 15: The Universe

The Universe known to us now!

Our galaxy, the Milky Way, is just one of millions of other galaxies. It is estimated that there are no less than 100,000 million galaxies in our universe. Some of these galaxies now known have 13 times more stars than our galaxy!

Just as stars gather in groups inside of our galaxy, galaxies also cluster together in groups.

For example, next to our galaxy are at least 24 other galaxies collectively considered our neighboring "local group." Within this local group, a galaxy called "Andromeda," which is about 2.2 million light-years away, is 1½ times larger than our galaxy.

The collection or group that our galaxy belongs to extends many millions of light-years! In this group there are also areas that are more concentrated with galaxies than others. For example, there is a place called "VERGO" that has a mass of about 800 thousand million times that of the sun, with a width of about 150,000 light-years!

In another collection of galaxies, there is another gathering called the "COMA," which is 450 thousand million light-years away. This collection is no less than 10 million light-years wide! It has even more galaxies than the collection to which our galaxy belongs.

All of these galaxies are moving away from us at an incredible speed. Some of these galaxies project through space at speeds of around 288,000 kilometers a second! So the universe is expanding and stretching!

Even with such great speeds, unimaginable sizes, and astronomical number of galaxies, the probability of two galaxies colliding is similar to that of two flies crashing into one another—one being released in Paris and the other in Washington, DC! That is because "space" is so much larger.

To where do these galaxies travel? What is this "space" that contains and envelops them?

I believe that the human mind's ability to even imagine this has ceased a long time ago!

Conclusion:
Return to our "home"!

The journey is over, and we have returned back to our planet Earth!

I was so amazed!

I am "nothing" in comparison to "Earth," "Earth" is "not anything" comparing to our "solar system," "the solar system" is "nil" in the face of the "galaxy," and "the Milky Way" is "zero" in front of the known "universe"!

Unimaginable masses, unthinkable distances, and inconceivable "energy"!

What is the "source" of "all" of this?

What is the "origin" of all this "harmony and incredible assembly"?

What is the "power" that controls these "mind blowing structures"?

What is the "intelligence" that brings together all these "shocking wonders"?

This Book is an open invitation for everyone to think genuinely and find answers to the above questions, and to open his or her heart to accept the truth!

CPSIA information can be obtained at www.ICGtesting.com
Printed in the USA
BVOW04s2301040514

352418BV00015B/60/P